"Look! A contest!" Sue said.

"Can you play marbles?"
asked Harry.

"Not yet," said Sue.

1

Sue's uncle came over
for dinner.

"I'll teach you to play
marbles," Uncle Hoyt said.
"This blue marble might
help you win."

Uncle Hoyt hit some
marbles for Sue. "You must
use a fast motion," he said.

It was Sue's turn to play.
"I'll use my blue marble,"
she said.

Sue hit a marble with
a fast action. "I can play this
game," said Sue.

4

Sue played marbles for
three weeks. At night her
hands hurt. Sue put some
lotion on them.

It was the day of the marble contest. Where was Sue's blue marble? "I must have my blue marble to play," Sue said.

Sue's portion of marbles was not that big.

"Where could my blue marble be?" she said sadly.

"Maybe you don't need your blue marble today," said her mom.

Sue's mother was right. Sue did not need her blue marble to play.

Sue's fast actions helped her win the contest. "Uncle Hoyt helped too," said Sue.

The End